Unveiling the Secrets: The Truth About Genetically Modified Food

Copyright Page

TITLE: Unveiling the Secrets: The Truth About Genetically Modified Food

1ST Edition

ISBN: 9798223951209

Table of Contents

Unveiling the Secrets – The Truth About Genetically Modified Foods

By Roberto Miguel Rodriguez

Chapter 1: The Truth About Genetically Modified Food

The History and Development of Genetically Modified Food

Genetically Modified Food (GM food) is a topic that has gained significant public attention in recent years. As consumers become more concerned about the origins and composition of their food, it is important to understand the history and development of genetically modified organisms (GMOs) and their impact on various aspects of society.

The concept of genetic modification dates back to ancient times when farmers selectively bred plants and animals to enhance desirable traits. However, the modern era of GMOs began in the 1970s with the advent of recombinant DNA technology. This breakthrough allowed scientists to transfer genes between different species, creating organisms with new characteristics.

One of the first genetically modified crops to be commercialized was the Flavr Savr tomato, which hit the market in 1994. This tomato was engineered to have a longer shelf life by suppressing the gene responsible for ripening. Since then, a wide range of GM crops have been developed, including corn, soybeans, and cotton. These crops are often modified to be resistant to pests, diseases, or herbicides, resulting in increased yields and reduced use of chemicals.

Despite the potential benefits of GM food, there are concerns about its impact on human health, particularly in children. Some studies suggest that GMOs may trigger allergic reactions or have long-term health effects, although the scientific consensus is that GM food is safe for consumption. Nevertheless, more research is needed to fully understand the health implications, especially for vulnerable populations such as children.

The environmental impact of GM food production is another important consideration. While GM crops can reduce the need for chemical pesticides and herbicides, they may also have unintended effects on ecosystems. For example, the widespread use of herbicide-resistant crops has led to the emergence of herbicide-resistant weeds, requiring even more potent chemicals to control them.

Ethical concerns surrounding GM food revolve around issues of corporate control and the potential for genetic contamination of organic or traditional crops. Critics argue that the patenting of GM seeds by multinational corporations restricts farmers' freedom to save and exchange seeds, perpetuating a cycle of dependency. Additionally, there are fears that GM crops can crossbreed with wild relatives, leading to the spread of modified genes into natural ecosystems.

From an economic perspective, GM food has the potential to alleviate food insecurity and malnutrition in developing countries. Drought-tolerant or vitamin-enriched crops could provide a sustainable solution to nutritional deficiencies. However, the high cost of GM seeds and the dependence on multinational corporations for seed supply can create economic disparities and hinder small-scale farmers' access to these technologies.

Regulation and labeling of GM food vary across countries. Some nations have strict regulations and mandatory labeling, while others have more lenient policies. This lack of harmonization makes it difficult for consumers to make informed choices and for scientists to conduct comprehensive studies on the long-term effects of GM food.

In conclusion, the history and development of genetically modified food are complex and multifaceted. While GM food has the potential to address various challenges, such as food security and malnutrition, it also raises concerns about health, the environment, ethics, and economic disparities. It is crucial for the public to be aware of these issues and engage in informed discussions to shape the future of GM food.

The Science Behind Genetic Modification

In order to understand the truth about genetically modified food, it is important to delve into the science behind genetic modification. Genetic modification, also known as genetic engineering or biotechnology, involves the deliberate manipulation of an organism's genetic material to achieve desired traits or characteristics. This process allows scientists to transfer specific genes from one organism to another, including across different species.

The health effects of genetically modified food on children have been a subject of concern for many parents. Numerous studies have been conducted to assess the safety of genetically modified food consumption. The consensus among scientific and regulatory bodies, such as the World Health Organization and the National Academy of Sciences, is that genetically modified foods currently on the market are safe for human consumption. However, ongoing research and monitoring are necessary to ensure their continued safety.

Genetic modification also has environmental implications. The production of genetically modified crops often involves the use of herbicides and pesticides, which can have detrimental effects on ecosystems. On the other hand, genetic modification has the potential to reduce the need for chemical inputs by creating crops that are resistant to pests and diseases. This can lead to a more sustainable and environmentally friendly agricultural system.

Ethical considerations surround the use of genetic modification in food production. Some argue that it is unnatural and goes against the principles of natural selection. Others raise concerns about the potential for corporate control over the food supply and the impact on small farmers. These ethical dilemmas must be carefully considered and debated in order to make informed decisions about the use of genetic modification in food production.

The economic implications of genetically modified food in developing countries are complex. While genetically modified crops have the potential to increase yields and reduce losses due to pests and diseases, they can also lead to increased dependence on multinational corporations for seeds and other agricultural inputs. This can have both positive and negative impacts on small-scale farmers and rural communities.

Genetically modified food also has the potential to contribute to global food security. By improving crop yields and enhancing nutritional content, genetically modified crops can help alleviate hunger and malnutrition in developing countries. However, it is important to ensure that these benefits are accessible to those who need them most and that the potential risks are carefully managed.

In addition to addressing nutritional deficiencies, genetic modification can also have an impact on biodiversity. The introduction of genetically modified crops can potentially lead to the loss of native plant varieties and the disruption of natural ecosystems. Thus, careful consideration must be given to the potential environmental impacts of genetic modification.

Regulation and labeling of genetically modified food is a contentious issue. Many countries have implemented regulations to ensure the safety and proper labeling of genetically modified products. However, there is ongoing debate about the adequacy of these regulations and the need for more transparency in the food industry.

Furthermore, there is a growing body of evidence suggesting a link between genetically modified food and food allergies. Some studies have shown that genetically modified crops can express new proteins that may trigger allergic reactions in susceptible individuals. This highlights the importance of rigorous testing and monitoring of genetically modified crops to ensure their safety for those with food allergies.

Finally, genetically modified crops have the potential to resist pests and diseases, reducing the need for chemical inputs. This can lead to more sustainable and environmentally friendly agricultural practices. By reducing the use of pesticides and herbicides, genetically modified crops can help protect biodiversity and promote a healthier ecosystem.

In conclusion, understanding the science behind genetic modification is crucial to forming an informed opinion about genetically modified food. It is important to consider the health effects, environmental impact, ethical considerations, economic implications, and regulatory aspects of genetically modified food. By weighing these factors, we can make decisions that promote the safe and responsible use of genetic modification in food production.

The Benefits and Risks of Genetically Modified Food

Genetically Modified (GM) food has been a topic of debate for years, with proponents and critics voicing their opinions regarding its safety and potential impact on various aspects of our lives. In this subchapter, we will delve into the benefits and risks associated with GM food, addressing important concerns and shedding light on the truth behind this controversial topic.

First and foremost, it is crucial to acknowledge that GM food offers several potential benefits. One of the key advantages is increased crop yield, which can help combat food scarcity and contribute to global food security. GM crops are engineered to be resistant to pests, diseases, and adverse weather conditions, ensuring higher productivity and reducing crop losses. Additionally, genetically modified crops can be enriched with essential nutrients, helping to combat nutritional deficiencies that are prevalent in many developing countries.

However, it is essential to consider the potential risks associated with GM food. One concern is the potential impact on human health, particularly in children. Critics argue that consuming GM food may lead to allergies or other adverse health effects. While extensive research and testing have been conducted to ensure the safety of GM food, more studies are needed to fully understand any potential long-term effects.

Furthermore, the environmental impact of GM food production is a significant consideration. The use of genetically modified crops has the potential to reduce the need for chemical pesticides and herbicides, thus minimizing their negative impact on the environment. However, there are concerns about cross-pollination with non-GM crops, which could lead to the loss of biodiversity and the creation of superweeds.

Ethical considerations are also important when discussing GM food. Critics argue that patenting GM seeds and monopolizing the market can have detrimental effects on small-scale farmers, particularly in developing countries. The economic implications of GM food in these regions must be carefully examined to ensure fair and equitable distribution of resources.

In terms of regulation and labeling, there is an ongoing debate on whether GM food should be more strictly regulated and labeled to provide consumers with informed choices. Some argue that greater transparency is necessary to address concerns about potential health risks and to respect individuals' right to know what they are consuming.

In conclusion, the benefits and risks of genetically modified food are complex and multifaceted. While GM food has the potential to address global food security, combat nutritional deficiencies, and reduce environmental impact, it also raises concerns about human health, biodiversity, and ethical considerations. It is essential for the public to be well-informed about these issues to make educated decisions and actively participate in shaping the future of our food system.

The Controversy Surrounding Genetically Modified Food

Introduction:

Genetically Modified Food (GMF) has been a subject of intense debate and controversy for several years. As our understanding of genetic engineering progresses, so does the concern surrounding the potential risks and benefits of this technology. This subchapter aims to shed light on the various aspects of this controversy, addressing the concerns of the public regarding the truth about genetically modified food.

The Truth About Genetically Modified Food:

There is a growing skepticism among the public regarding the safety of genetically modified food. Many individuals question whether these foods are truly safe for consumption and if long-term health effects have been adequately studied. This section will explore the scientific evidence surrounding GMF safety and dispel any misconceptions.

The Health Effects of Genetically Modified Food on Children:

Children, being more vulnerable to potential risks, require special attention when it comes to evaluating the impact of GMF on their health. This section will discuss the available research on the effects of GMF consumption on children, exploring any potential risks or benefits that may arise.

The Environmental Impact of Genetically Modified Food Production:

GMF production methods can have significant environmental implications. This section will examine the environmental impact of genetically modified crops, including their potential effects on soil health, water quality, and biodiversity. It will also consider the sustainability of GMF production in the long term.

The Ethical Considerations of Genetically Modified Food:

The development and use of genetically modified food raise ethical questions related to human autonomy, environmental justice, and the rights of future generations. This section will delve into these ethical considerations and explore the different perspectives surrounding them.

The Economic Implications of Genetically Modified Food in Developing Countries:

GMF has the potential to alleviate hunger and poverty in developing countries, but it also raises concerns about the control of seed resources by multinational corporations. This section will analyze the economic implications of GMF in developing countries, including the potential benefits and challenges faced by local farmers.

The Role of Genetically Modified Food in Global Food Security:

With a growing global population, ensuring food security is crucial. This section will discuss the potential role of GMF in addressing food shortages and increasing agricultural productivity, while also considering the social, economic, and environmental consequences.

Genetically Modified Food and Its Potential to Combat Nutritional Deficiencies:

GMF has the potential to address nutritional deficiencies by enhancing the nutritional content of crops. This section will explore the promise and challenges of genetically modified crops in combating malnutrition and improving public health.

The Impact of Genetically Modified Food on Biodiversity:

Introducing genetically modified crops into the environment can have unintended consequences on biodiversity. This section will examine the potential impact of GMF on biodiversity, including the risks to native species and the importance of preserving genetic diversity.

The Regulation and Labeling of Genetically Modified Food:

Regulation and labeling of GMF are essential to inform consumers and ensure their right to choose. This section will discuss the current regulatory frameworks and labeling practices, exploring the need for transparency and clear information for consumers.

Genetically Modified Food and Its Connection to Food Allergies:

There are concerns that genetically modified crops may trigger allergic reactions in some individuals. This section will examine the available evidence on the connection between GMF and food allergies, highlighting the importance of proper allergenicity testing and labeling.

Genetically Modified Crops and Their Resistance to Pests and Diseases:

One of the potential benefits of GMF is their ability to resist pests and diseases, reducing the need for chemical pesticides. This section will explore the effectiveness and sustainability of genetically modified crops in pest and disease management, considering both the benefits and potential drawbacks.

Conclusion:

The controversy surrounding genetically modified food is multifaceted and involves various stakeholders. By examining the different aspects of this controversy, this subchapter aims to provide the public with a comprehensive understanding of the truth about genetically modified food and enable them to make informed decisions about its consumption.

Chapter 2: The Health Effects of Genetically Modified Food on Children

The Impact of Genetically Modified Food on Child Development

In recent years, the topic of genetically modified (GM) food has sparked controversy and raised concerns among the public. As parents, it is crucial to understand the potential impact of these foods on child development. This subchapter aims to shed light on the truth about genetically modified food and its effects on our children's health and overall well-being.

Numerous studies have been conducted to determine the health effects of genetically modified food on children. Some research suggests that these foods may have adverse effects on child development, including potential risks to their immune system and gut microbiome. As children are still growing and developing, their delicate systems may be more susceptible to the potential negative impacts of genetically modified organisms (GMOs).

Additionally, the environmental impact of genetically modified food production should not be overlooked. The excessive use of pesticides and herbicides associated with GM crops can lead to environmental contamination and pose risks to children's health. Exposure to these harmful chemicals has been linked to developmental delays and behavioral disorders.

Ethical considerations also come into play when discussing genetically modified food. Many argue that altering the genetic makeup of organisms goes against the natural order and raises ethical concerns regarding the long-term consequences for both humans and the environment. It is essential to weigh the potential benefits against the ethical implications when considering the use of genetically modified food.

Furthermore, the economic implications of genetically modified food in developing countries should be taken into account. While proponents argue that GM crops can increase yields and food production, critics argue that they may lead to dependency on large biotech companies. This can have significant socioeconomic impacts, affecting access to nutritious food for vulnerable populations.

Addressing global food security is another critical aspect of the genetically modified food debate. Proponents argue that GM crops have the potential to combat nutritional deficiencies and alleviate hunger worldwide. However, critics raise concerns about the long-term sustainability and potential consequences of relying heavily on genetically modified food.

Considering the potential impact on biodiversity, it is important to evaluate the effects of genetically modified crops on the environment and natural ecosystems. The introduction of GM crops can disrupt native species and threaten biodiversity, which has far-reaching consequences for ecosystems and the overall health of the planet.

Regulation and labeling of genetically modified food are also important considerations. Many argue that consumers have the right to know whether the food they are purchasing contains GMOs. Proper labeling and transparent regulation are crucial for informed decision-making and allowing individuals to make choices that align with their values and health concerns.

Moreover, the connection between genetically modified food and food allergies has raised concerns among parents. Some studies suggest that the introduction of GM crops may contribute to the increased prevalence of food allergies in children. It is essential to understand the potential risks and ensure the safety of genetically modified food for those with allergies.

Lastly, genetically modified crops' resistance to pests and diseases has been touted as a significant advantage. However, the long-term consequences of this genetic modification on the environment and human health need to be carefully evaluated. The potential for pests and diseases to develop resistance to genetically modified crops is a concern that must be addressed.

In conclusion, understanding the impact of genetically modified food on child development is crucial for parents and the public. The health effects, environmental impact, ethical considerations, economic implications, and global food security all play a role in shaping our understanding of genetically modified food. It is important to consider these factors when making informed decisions about the food we feed our children and the future of our global food system.

Potential Allergenic Reactions in Children

When it comes to the topic of genetically modified (GM) food, one aspect that cannot be overlooked is the potential allergenic reactions it may pose, especially in children. As concerned parents, it is important to be aware of the potential risks associated with consuming genetically modified organisms (GMOs) and how they may impact the health of our little ones.

While proponents of GM food argue that they have undergone rigorous testing to ensure their safety, there are still concerns about the introduction of new allergens into the food supply. Genetic modifications can introduce proteins or other substances that are not typically found in nature, potentially triggering allergic reactions in susceptible individuals.

Children, in particular, are more vulnerable to developing allergies due to their still developing immune systems. Studies have shown that certain genetically modified crops, such as soybeans, corn, and peanuts, have been modified to be more resistant to pests or herbicides, which may inadvertently increase the levels of allergenic proteins in these crops. This can pose a significant risk to children who already have allergies or are predisposed to developing them.

Moreover, the lack of mandatory labeling of GM food makes it challenging for parents to identify and avoid products that may contain potential allergens. This lack of transparency further adds to the concerns surrounding the safety and potential allergenicity of GM food.

It is crucial that parents educate themselves about the potential risks and make informed choices regarding the food they provide for their children. Opting for organic or non-GMO products can provide an alternative and potentially safer option. Additionally, consulting with healthcare professionals can help determine if a child is at a higher risk of developing allergies and guide parents in making appropriate dietary choices.

In conclusion, the potential allergenic reactions in children cannot be overlooked when discussing the truth about genetically modified food. As parents, it is our responsibility to stay informed and make conscious decisions to ensure the health and well-being of our children. By understanding the potential risks associated with GM food and taking appropriate measures to avoid potential allergens, we can help protect our children from the potential adverse effects of genetically modified organisms.

Long-term Health Consequences of Consuming Genetically Modified Food

Genetically Modified (GM) food has become increasingly prevalent in our modern society, but what are the potential long-term health consequences of consuming these genetically altered products? In this subchapter, we will explore the various health implications associated with GM food consumption, shedding light on the truth behind this controversial topic.

Several studies have shown that consuming GM food can have adverse effects on human health. One concern is the potential for allergenic reactions. Genes from one species are often transferred into another during the genetic modification process, which may introduce new allergens into the food supply. This poses a significant risk to individuals with existing food allergies, as well as to those who have never experienced allergic reactions before.

Another long-term health consequence of consuming GM food is the development of antibiotic resistance. Genetic engineering techniques often involve the use of antibiotic resistance genes as markers, leaving the possibility of transferring these genes to bacteria in the human gut. This could lead to the emergence of antibiotic-resistant strains of bacteria, making it more challenging to treat bacterial infections effectively.

Moreover, the impact of GM food on the human microbiome is still not fully understood. The microbiome plays a crucial role in human health, influencing various physiological processes, such as digestion and immune function. Consuming GM food may disrupt the delicate balance of bacteria in the gut, potentially leading to long-term health issues.

Furthermore, some studies suggest a potential link between GM food consumption and chronic diseases such as cancer and obesity. Although more research is needed to establish a definitive connection, the presence of toxins and changes in nutrient composition in GM food could contribute to these health concerns.

It is essential for the public to be aware of these potential long-term health consequences and to demand further research and transparent labeling of GM food products. As consumers, we have the right to make informed choices about the foods we consume, taking into account the potential risks and benefits associated with GM food.

In conclusion, the long-term health consequences of consuming genetically modified food are a matter of concern. Allergenic reactions, antibiotic resistance, disruption of the microbiome, and potential links to chronic diseases are among the issues that need further investigation. It is crucial for the public to stay informed and engaged in the ongoing debate surrounding GM food, as our health and well-being may depend on it.

Chapter 3: The Environmental Impact of Genetically Modified Food Production

Effects on Soil Quality and Ecosystems

Genetically Modified (GM) food production has had a significant impact on soil quality and ecosystems. While proponents argue that GM crops increase yields and reduce the need for pesticides, critics are concerned about the long-term effects on our environment.

One of the main concerns is the potential for increased pesticide use. GM crops are often engineered to be resistant to pests, allowing farmers to use fewer chemical pesticides. However, this can lead to the development of pesticide-resistant pests, creating a cycle of increased pesticide use. This not only harms the soil quality but also poses a threat to beneficial insects and other organisms in the ecosystem.

Additionally, the introduction of GM crops can lead to the loss of biodiversity. As GM crops are often engineered to be resistant to pests and diseases, they can outcompete native plant species, reducing the diversity of plant life in the ecosystem. This loss of biodiversity can have ripple effects on other organisms, including pollinators and other wildlife that rely on diverse plant species for food and habitat.

Moreover, the use of GM crops can also have unintended consequences for soil health. For example, some GM crops are engineered to produce their own pesticides, such as Bacillus thuringiensis (Bt) toxin. While these toxins are effective against pests, they can also harm beneficial soil organisms, including earthworms and beneficial bacteria. These organisms play a crucial role in maintaining soil fertility and nutrient cycling, so their decline can have long-term consequences for soil health.

Furthermore, the cultivation of GM crops often requires intensive tillage practices, which can lead to increased erosion and soil degradation. Tillage breaks up soil structure and exposes it to erosion by wind and water. This can result in the loss of topsoil, which is rich in organic matter and essential nutrients for plant growth. The loss of topsoil not only reduces the productivity of the land but also contributes to sedimentation in water bodies, leading to water pollution and further environmental damage.

In conclusion, the widespread adoption of GM crops has had significant effects on soil quality and ecosystems. While proponents argue that GM crops can increase yields and reduce pesticide use, concerns remain regarding the long-term impact on biodiversity, soil health, and the environment. It is crucial for policymakers, scientists, and the public to carefully evaluate the risks and benefits of GM food production to ensure the sustainability of our ecosystems and the health of our soil.

Genetic Contamination of Wild Plants

In the world of genetically modified (GM) food, one of the most concerning issues is the genetic contamination of wild plants. This phenomenon occurs when the genes from genetically modified crops are unintentionally transferred to their wild relatives or other related plant species. The consequences of this genetic contamination can have far-reaching effects on various aspects, including the environment, biodiversity, and even human health.

Genetic contamination poses a significant threat to the integrity of natural ecosystems. Wild plants play a vital role in maintaining the balance of biodiversity and ecological stability. When genetically modified genes infiltrate wild populations, they can disrupt the natural genetic diversity and potentially lead to the extinction of native plant species. As these wild plants serve as a habitat and food source for countless organisms, the loss of their genetic diversity can have cascading effects throughout the entire ecosystem.

Moreover, the health effects of genetically modified food on children cannot be overlooked. Children are more vulnerable to potential risks associated with GM crops due to their developing immune systems and higher susceptibility to allergies. The unintentional consumption of genetically modified plants or their byproducts may have adverse effects on their health, including allergic reactions, digestive problems, and other yet unknown long-term consequences.

From an ethical standpoint, genetic contamination raises concerns about the unintended consequences of manipulating nature. Genetic modification often involves the use of synthetic chemicals and DNA manipulation techniques that may have unforeseen effects on the environment and human health. These ethical considerations call for a careful examination of the risks and benefits associated with genetically modified food production and consumption.

Furthermore, the economic implications of genetic contamination in developing countries cannot be ignored. Many developing nations heavily rely on agriculture for their livelihoods and export markets. The presence of genetically modified genes in their crops can lead to trade barriers and economic losses. Developing countries must carefully weigh the potential benefits of adopting genetically modified crops against the risks of genetic contamination and its impact on their agricultural industries.

Addressing the issue of genetic contamination requires effective regulation and labeling of genetically modified food. Clear and transparent labeling practices enable consumers to make informed choices and avoid potential health risks associated with genetically modified products. Additionally, strict regulations should be in place to prevent the unintentional release and spread of genetically modified genes into the environment.

In conclusion, the genetic contamination of wild plants is a critical issue that must be carefully examined and addressed. It affects various aspects, including the environment, human health, biodiversity, and economic stability. As the public, it is essential to stay informed about the potential risks and benefits of genetically modified food and actively engage in discussions and decision-making processes regarding its production, regulation, and consumption.

Impact on Non-target Organisms and Biodiversity

The introduction of genetically modified (GM) food has raised concerns about its potential impact on non-target organisms and biodiversity. Non-target organisms refer to those species that are not the intended target of the GM crop but may be affected by its cultivation or consumption.

One major concern is the potential harm to beneficial insects such as bees and butterflies. These insects play a crucial role in pollination, which is essential for the reproduction of many plant species. Some studies have shown that certain GM crops, particularly those engineered to produce insecticides, can have adverse effects on beneficial insects. For example, the pollen from GM crops may contain toxins that can be harmful or even lethal to these insects.

Furthermore, the cultivation of GM crops can also have unintended consequences on other organisms in the ecosystem. For instance, the use of herbicide-tolerant GM crops may lead to the increased use of herbicides, which can harm non-target plants and disrupt the balance of biodiversity in agricultural landscapes. Additionally, the transfer of GM traits to wild relatives of cultivated plants through cross-pollination has raised concerns about the potential for genetic contamination and the loss of genetic diversity in wild populations.

Biodiversity is essential for the stability and resilience of ecosystems. It provides a wide range of ecosystem services, including nutrient cycling, pest control, and soil fertility. The impact of GM crops on biodiversity can vary depending on several factors, including the specific trait introduced, the crop species, and the local environment.

To address these concerns, regulatory frameworks have been established in many countries to assess the potential environmental impacts of GM crops before they are approved for commercial cultivation. These regulations require rigorous testing of GM crops to evaluate their effects on non-target organisms and biodiversity.

However, it is important to note that the potential impact of GM crops on non-target organisms and biodiversity is a complex and multifaceted issue. Scientific research and ongoing monitoring are crucial to better understand the long-term effects of GM crops on ecosystems. Additionally, the adoption of sustainable agricultural practices that minimize the use of pesticides and promote biodiversity conservation can help mitigate any potential negative impacts of GM crops on non-target organisms and biodiversity.

In conclusion, the introduction of GM food has raised concerns about its impact on non-target organisms and biodiversity. While some studies have shown potential adverse effects on beneficial insects and unintended consequences on ecosystem balance, regulatory frameworks and ongoing research aim to address these concerns and ensure the safe and sustainable use of GM crops.

Chapter 4: The Ethical Considerations of Genetically Modified Food

Ownership and Control of Genetic Resources

In the realm of genetically modified food, the issue of ownership and control of genetic resources is a crucial aspect that needs to be addressed. Genetic resources refer to the raw materials, such as plants, animals, and microorganisms, that possess genetic information that can be utilized for various purposes, including the development of genetically modified organisms (GMOs). This subchapter unveils the truth about the ownership and control of genetic resources in the context of genetically modified food.

One of the major concerns surrounding genetic resources is the appropriation of these resources by corporations and developed nations, particularly from developing countries. Many developing nations possess rich biodiversity and genetic resources that have immense potential for the development of GMOs. However, due to various factors, including lack of awareness, technology, and resources, they often lack the capacity to utilize these resources themselves.

This has led to a situation where multinational corporations and developed countries have gained control over the genetic resources of developing nations. They have exploited these resources to develop GMOs that are then patented and owned by these entities. This raises ethical concerns as it jeopardizes the rights of indigenous communities and traditional farmers who have been the custodians of these genetic resources for generations.

Moreover, the economic implications of this control are significant. Developing countries, which are often dependent on agriculture for their livelihoods, face challenges in accessing and utilizing genetically modified seeds due to intellectual property rights and high costs of licensing. This further perpetuates the cycle of dependency on developed nations and corporations.

Additionally, the control of genetic resources impacts biodiversity and environmental sustainability. The focus on a few genetically modified crops leads to a reduction in crop diversity, which in turn increases the vulnerability of food systems to pests, diseases, and climate change. Furthermore, the environmental impact of genetically modified food production, such as the use of pesticides and herbicides, needs to be considered in the context of ownership and control.

In order to address these issues, there is a need for international agreements and policies that safeguard the rights of developing nations and indigenous communities over their genetic resources. This includes equitable benefit-sharing mechanisms and the recognition of traditional knowledge. Furthermore, more research and investment should be directed towards building the capacity of developing countries to utilize their own genetic resources.

In conclusion, the ownership and control of genetic resources in the context of genetically modified food is a complex issue with far-reaching implications. It is crucial for the public to be aware of these issues and engage in discussions surrounding the ethical, economic, and environmental aspects. By understanding the truth about ownership and control, we can strive for a more equitable and sustainable future in the realm of genetically modified food.

Consent and Consumer Choice

In the realm of genetically modified (GM) food, consent and consumer choice play a crucial role. As members of the public, it is vital for us to understand the implications of consuming GM food and have the freedom to make informed decisions about what we put into our bodies.

The Truth About Genetically Modified Food

To begin, let's unveil the truth about genetically modified food. GM food refers to crops or animals that have undergone genetic engineering, where scientists modify their DNA to enhance certain traits. This technology has been widely used to increase crop yields, improve resistance to pests, and enhance nutritional content.

The Health Effects of Genetically Modified Food on Children

One niche that demands our attention is the health effects of genetically modified food on children. As parents, we want to ensure that our children's health is not compromised. Studies have shown conflicting evidence regarding the impact of GM food on human health. It is crucial to explore the long-term effects and potential risks associated with these products, especially when considering their consumption by the younger population.

The Environmental Impact of Genetically Modified Food Production

Another niche to consider is the environmental impact of genetically modified food production. While GM crops may offer benefits such as reduced pesticide use and increased productivity, their cultivation can also have unintended consequences. These include the potential for cross-pollination with wild relatives, the development of superweeds, and the disruption of natural ecosystems. Understanding and addressing these concerns is vital for sustainable agricultural practices.

The Ethical Considerations of Genetically Modified Food

Ethical considerations surrounding genetically modified food cannot be ignored. Questions arise about the ownership and control of GM seeds, the potential for corporate dominance in the food industry, and the impact on small-scale farmers. It is crucial to navigate these ethical complexities and ensure that the development and distribution of GM food are guided by fair and just principles.

The Economic Implications of Genetically Modified Food in Developing Countries

In developing countries, the economic implications of genetically modified food are significant. While some argue that GM crops can help alleviate poverty and improve food security, others raise concerns about increased dependency on multinational corporations and the potential displacement of local farmers. Evaluating the economic impact of GM food in developing countries is essential to strike a balance between progress and sustainability.

These are just a few of the many aspects that highlight the importance of consent and consumer choice in the context of genetically modified food. By understanding the truth, evaluating health effects, considering environmental impacts, addressing ethical concerns, and exploring economic implications, we can make informed decisions about our food choices. It is essential to continue research, engage in dialogue, and advocate for transparent regulation and labeling to ensure that our consent is respected and our choices are meaningful.

Animal Welfare Concerns

When discussing the topic of genetically modified (GM) food, it is crucial to address the animal welfare concerns that arise from its production and consumption. While the focus is often on the potential health effects and environmental impact, the ethical considerations regarding animal welfare cannot be overlooked.

One of the primary concerns is the use of animals in GM food research and testing. Animals are often used to study the safety and efficacy of genetically modified organisms (GMOs) before they are deemed fit for human consumption. This raises questions about the welfare of these animals and whether their use is necessary or ethical.

Another issue is the impact of GM crops on wildlife and biodiversity. GM crops are designed to resist pests and diseases, often through the use of pesticides or herbicides. However, this can have unintended consequences for non-target organisms, such as beneficial insects or birds that rely on these organisms for food. The alteration of natural ecosystems through the introduction of GM crops can disrupt the delicate balance of biodiversity.

Furthermore, the use of GM feed in animal agriculture raises concerns about animal welfare. Many livestock animals are fed GM crops, such as corn or soybeans, which have been genetically modified for increased productivity or resistance to herbicides. The long-term effects of consuming GM feed on animal health and welfare are not yet fully understood.

In addition, there is a debate about the ethics of genetically modifying animals themselves. Some GM animals have been created for various purposes, such as producing pharmaceuticals or improving agricultural productivity. However, altering the genetic makeup of animals raises ethical questions about their welfare, as well as the potential unintended consequences for ecosystems if these animals were to escape into the wild.

Overall, animal welfare concerns are an important aspect of the broader discussion about genetically modified food. As we continue to explore the potential benefits and risks of GM technology, it is crucial that we consider the ethical implications and strive for a balanced approach that respects both human and animal well-being.

Chapter 5: The Economic Implications of Genetically Modified Food in Developing Countries

Impact on Traditional Farming Practices

The advent of genetically modified food has had a significant impact on traditional farming practices around the world. Traditional farming methods, which have been passed down through generations, are now being challenged by the introduction of genetically modified crops.

One of the major impacts is the increased use of pesticides and herbicides. Genetically modified crops are often engineered to be resistant to certain pests or diseases, which means farmers can use fewer chemicals to protect their crops. This has led to a reduction in the use of traditional methods such as crop rotation and companion planting, which were used to naturally control pests and maintain soil health.

Furthermore, the introduction of genetically modified crops has also caused a shift towards monoculture farming. Traditional farming practices emphasized diversity in crops, allowing for a more balanced ecosystem and reducing the risk of crop failure. However, genetically modified crops are often designed to be high-yielding and uniform, leading to a decrease in crop diversity. This can have negative consequences for biodiversity and the resilience of farming systems.

The economic implications of genetically modified food on traditional farming practices cannot be ignored. In many developing countries, small-scale farmers rely on traditional farming methods for their livelihoods. However, the introduction of genetically modified crops has often led to the displacement of these farmers, as they struggle to compete with large-scale, genetically modified crop producers. This has resulted in the loss of traditional knowledge and cultural practices associated with farming.

Moreover, the regulation and labeling of genetically modified food have also impacted traditional farming practices. Many countries have implemented regulations to ensure the safety and proper labeling of genetically modified products. This has increased the administrative burden on traditional farmers, who may not have the resources or knowledge to comply with these regulations.

In conclusion, the impact of genetically modified food on traditional farming practices is multi-faceted. While it has brought about certain benefits such as reduced pesticide use and increased crop yields, it has also led to the loss of traditional knowledge, decreased biodiversity, and economic challenges for small-scale farmers. As the debate surrounding genetically modified food continues, it is important to consider the long-term implications for traditional farming practices and the sustainability of our food systems.

Access to Seeds and Intellectual Property Rights

One of the key issues surrounding genetically modified (GM) food is the issue of access to seeds and intellectual property rights. This subchapter aims to shed light on this topic, exploring the implications it has on various aspects of society.

When it comes to access to seeds, it is important to note that a handful of multinational corporations dominate the global seed market. These corporations have developed patented GM seeds, which means that farmers are required to purchase new seeds every planting season, rather than saving and replanting seeds from their harvests. This has significant economic implications, particularly for small-scale farmers in developing countries who may struggle to afford these expensive seeds.

Intellectual property rights play a crucial role in this dynamic. Patents allow seed companies to protect their investments in research and development, but they also restrict farmers' ability to save and replant seeds. This has raised concerns about the concentration of power in the hands of a few corporations, as well as the potential for seed dependency in the long run.

From an ethical standpoint, the issue of access to seeds raises questions about food sovereignty and the rights of farmers. Many argue that farmers should have the freedom to save and exchange seeds, as they have done for centuries, rather than being reliant on patented seeds controlled by multinational corporations. This also raises concerns about the impact on biodiversity, as the cultivation of a limited number of patented GM crops may lead to a loss of traditional crop varieties.

The regulation and labeling of GM food also intersects with the issue of access to seeds and intellectual property rights. Some argue that labeling should be mandatory to ensure consumer choice and transparency. However, others argue that labeling may stigmatize GM food, potentially impacting its marketability and the livelihoods of farmers who grow these crops.

In conclusion, access to seeds and intellectual property rights are complex issues with far-reaching implications. They touch on various aspects of society, including the economic, ethical, environmental, and health-related aspects of GM food. As the debate continues, it is crucial for the public to be informed and engaged in discussions surrounding these issues to ensure a balanced and fair approach to the production and distribution of GM food.

Economic Disparities and Dependence on Multinational Corporations

In the global food market, economic disparities and the dependence on multinational corporations are critical issues that need to be addressed when discussing genetically modified food (GMF). This subchapter aims to shed light on the economic implications of GMF, particularly in developing countries, and how multinational corporations play a significant role in this complex landscape.

Developing countries often face economic hardships and struggle with food security. The introduction of GMF has the potential to address some of these challenges by increasing crop yields, reducing production costs, and enhancing resistance to pests and diseases. However, the economic implications are far from straightforward.

Multinational corporations, with their vast resources and patented GMF technologies, often penetrate developing markets. While they promise economic benefits, such as increased agricultural productivity and job creation, the reality can be more complex. Many argue that the dominance of these corporations leads to economic disparities and a reliance on foreign entities. Local farmers may become dependent on these corporations for seeds, fertilizers, and pesticides, resulting in a loss of traditional farming practices and increased vulnerability to market fluctuations.

Additionally, the introduction of GMF can disrupt local economies and traditional agricultural systems. Small-scale farmers may struggle to compete with the large-scale production facilitated by multinational corporations. This can lead to a concentration of wealth and power in the hands of a few, exacerbating economic disparities within communities and countries.

It is essential to critically examine the relationship between multinational corporations, economic disparities, and GMF. By doing so, we can better understand the potential consequences and work towards more sustainable and inclusive agricultural systems. This includes supporting local farmers, promoting agricultural diversity, and ensuring that economic benefits are shared equitably.

In conclusion, economic disparities and the dependence on multinational corporations are significant factors in the GMF debate. Developing countries face complex challenges and potential benefits when it comes to adopting GMF technologies. It is crucial to approach these issues with a holistic perspective, considering the economic implications, social equity, and long-term sustainability. Only by doing so can we navigate the complexities of GMF and work towards a more inclusive and equitable global food system.

Chapter 6: The Role of Genetically Modified Food in Global Food Security

Increasing Crop Yield and Food Production

In order to tackle the global challenge of feeding a rapidly growing population, it is crucial to explore ways to increase crop yield and food production. Genetically modified food (GM food) has emerged as a potential solution to address this issue. By utilizing biotechnology techniques, scientists have been able to enhance the characteristics of crops, resulting in higher yield and improved resistance to pests and diseases.

One of the main benefits of GM food is its ability to increase crop yield. Through genetic modifications, scientists have been able to develop crops that are more resistant to environmental stressors such as drought, heat, and salinity. These crops have the potential to thrive in regions where traditional crops struggle to grow, thus increasing overall food production.

Furthermore, GM food has the potential to combat nutritional deficiencies. Scientists have successfully introduced genes into crops to enhance their nutritional content. For example, genetically modified rice, also known as Golden Rice, has been engineered to contain higher levels of vitamin A, which is crucial for preventing blindness and other health issues in developing countries where vitamin A deficiency is prevalent. This innovation holds great promise for addressing malnutrition and improving public health.

However, it is important to consider the potential environmental impact of GM food production. Critics argue that genetically modified crops may harm biodiversity and disrupt ecosystems. While there is ongoing debate on this topic, it is essential to ensure that the introduction of GM crops is done responsibly, with rigorous environmental risk assessments.

Ethical considerations also come into play when discussing GM food. Some individuals question the morality of altering the genetic makeup of organisms and the potential long-term consequences. These concerns highlight the need for careful regulation and transparent labeling of GM food products, allowing consumers to make informed choices.

From an economic standpoint, GM food has the potential to benefit developing countries by increasing agricultural productivity and improving food security. However, it is essential to ensure that these technologies are accessible and affordable for farmers in these regions, avoiding dependency on multinational corporations.

In conclusion, increasing crop yield and food production is a pressing global issue. Genetically modified food offers potential solutions to address this challenge, but it is important to carefully consider the health effects on children, the environmental impact, ethical considerations, economic implications, global food security, nutritional deficiencies, biodiversity, regulation and labeling, and the connection to food allergies. By addressing these concerns and implementing responsible practices, we can harness the benefits of genetically modified crops to meet the growing demand for food while ensuring the sustainability and well-being of our planet.

Drought and Disease Resistance

In recent years, the topic of genetically modified (GM) food has become a subject of intense debate and scrutiny. One of the key areas of interest is the potential for GM crops to exhibit resistance to drought and diseases. This subchapter aims to shed light on this important aspect of GM food, addressing the concerns and highlighting the benefits it can offer.

Drought is a major challenge faced by farmers worldwide, leading to significant crop losses and food insecurity. By incorporating specific genes into crops, scientists have been able to develop GM plants that exhibit enhanced drought tolerance. These genetically modified crops have the ability to survive and thrive under water-stressed conditions, thereby ensuring a more reliable food supply even in regions prone to drought. This advancement in crop technology has the potential to alleviate the suffering of millions of people facing hunger due to water scarcity.

Similarly, disease resistance is another critical factor in crop production. Traditional methods of disease control often involve the use of chemical pesticides, which can have detrimental effects on human health and the environment. Genetically modified crops, on the other hand, offer a sustainable and environmentally friendly solution. Through genetic engineering, scientists have successfully introduced genes that confer resistance to specific diseases, reducing the need for harmful chemicals and enhancing crop productivity. This not only benefits farmers but also reduces the risk of pesticide residues in food, ensuring safer and healthier produce for consumers.

While the potential benefits of drought and disease-resistant GM crops are evident, it is crucial to address the concerns raised by critics. Some argue that genetic modification may have unintended consequences, such as the creation of superweeds or the transfer of modified genes to non-target species. However, rigorous testing and regulatory frameworks are in place to assess the safety and potential risks associated with GM crops before they are approved for commercial use. Additionally, ongoing research and monitoring are conducted to ensure any potential environmental impacts are identified and addressed promptly.

In conclusion, the development of genetically modified crops with drought and disease resistance holds great promise for ensuring food security, reducing reliance on chemical pesticides, and promoting sustainable agriculture. However, it is essential to continue exploring the potential risks and benefits associated with GM technology to make informed decisions. Through transparent regulation, labeling, and public engagement, we can navigate the complexities of genetically modified food and harness its potential to address the pressing challenges of our time.

Challenges and Limitations for Developing Nations

In the global debate surrounding genetically modified (GM) food, it is crucial to acknowledge the unique challenges and limitations faced by developing nations. These countries, often grappling with poverty, food insecurity, and limited resources, must navigate the complex landscape of GM food production and consumption. Understanding the specific issues they face is essential in formulating effective policies and strategies.

One of the primary concerns is the economic implications of GM food in developing countries. While proponents argue that GM crops can increase yields and reduce production costs, the reality is more nuanced. Farmers in these nations often lack access to the necessary technology, resources, and knowledge required to adopt GM practices. This can result in increased dependency on multinational corporations that control the GM seed market, further exacerbating inequality and hindering the development of local agriculture.

Moreover, the health effects of genetically modified food, especially on children, pose a significant challenge. Limited research exists on the long-term impact of consuming GM products, and developing nations may lack the capacity to conduct comprehensive studies. Children, who are more vulnerable to potential health risks, could be disproportionately affected. It is essential to prioritize rigorous and independent research to ensure the safety of GM food consumption, especially in these vulnerable populations.

The environmental impact of GM food production is another critical concern. Developing countries often struggle with fragile ecosystems and limited regulatory frameworks. The introduction of GM crops can lead to unintended consequences, such as the loss of biodiversity and the emergence of pesticide-resistant pests. Balancing the potential benefits of increased productivity with the protection of local ecosystems is a complex task that requires careful consideration and sustainable practices.

Ethical considerations surrounding GM food also need to be addressed. Developing nations may face challenges in ensuring the equitable distribution of benefits and resources associated with GM crops. Additionally, the potential for genetic contamination of traditional crops and the loss of indigenous knowledge and seed diversity raise important ethical questions. Balancing the interests of local communities, multinational corporations, and global food security is a complex task that necessitates transparent and inclusive decision-making processes.

Furthermore, the regulation and labeling of GM food in developing countries can be inadequate or nonexistent. This lack of oversight can lead to potential health risks, consumer confusion, and the violation of international trade regulations. Strengthening regulatory frameworks and implementing clear labeling practices is crucial to protect consumers and ensure informed choices.

In conclusion, developing nations face unique challenges and limitations when it comes to genetically modified food. Economic implications, health effects, environmental impact, ethical considerations, and regulatory frameworks must all be carefully addressed to ensure the responsible adoption of GM technology. By taking into account the specific needs and contexts of these countries, we can strive for a more inclusive and sustainable approach to GM food production and consumption.

Chapter 7: Genetically Modified Food and Its Potential to Combat Nutritional Deficiencies

Biofortification and Enhanced Nutritional Content

In recent years, a new approach to addressing the global challenge of malnutrition has gained significant attention - biofortification. This subchapter delves into the concept of biofortification and its potential to enhance the nutritional content of genetically modified (GM) food.

Biofortification refers to the process of increasing the concentration of essential nutrients in food crops through conventional breeding or genetic engineering techniques. By enhancing the nutritional content of crops, biofortification aims to combat various micronutrient deficiencies, such as iron, zinc, and vitamin A, that affect millions of people worldwide.

One of the key advantages of biofortified GM crops is their ability to deliver essential nutrients directly to those who consume them. For example, Golden Rice, a genetically modified variety, has been engineered to contain beta-carotene, a precursor to vitamin A. This innovation has the potential to address vitamin A deficiency, which is a leading cause of blindness in many developing countries.

Furthermore, biofortified GM crops can play a significant role in improving the health of children. Studies have shown that children are particularly vulnerable to the negative health effects of malnutrition. By incorporating higher levels of essential nutrients into staple crops, biofortification could help combat the detrimental impact of nutrient deficiencies on child development and overall well-being.

While the environmental impact of GM food production is a legitimate concern, biofortification offers a more sustainable solution compared to traditional supplementation programs. Instead of relying on expensive and logistically challenging distribution of supplements, biofortification ensures that the nutrients are naturally present in the food consumed by communities, reducing the need for external interventions.

Ethical considerations surrounding GM food are also addressed through biofortification. By prioritizing the nutritional needs of vulnerable populations, biofortification represents a more ethically responsible approach to addressing malnutrition and promoting food security.

Biofortification also holds economic implications, especially for developing countries. By improving the nutritional content of staple crops, biofortification can reduce healthcare costs associated with malnutrition-related illnesses, enhance productivity, and alleviate poverty.

Moreover, the potential of biofortification to combat nutritional deficiencies goes hand-in-hand with the goal of achieving global food security. By enhancing the nutritional value of crops, biofortification contributes to a more sustainable and resilient food system, reducing the risk of food shortages and ensuring that communities have access to a diverse and nutritious diet.

In conclusion, biofortification and enhanced nutritional content in genetically modified food offer a promising solution to address the challenges of malnutrition and its wide-ranging impact on health, the environment, ethics, economics, and food security. By harnessing the potential of GM technology, biofortification paves the way for a future where crops can play a crucial role in eradicating nutrient deficiencies and improving the overall well-being of populations worldwide.

Addressing Vitamin and Mineral Deficiencies

One of the major concerns surrounding genetically modified (GM) food is its potential impact on human health and nutrition. In this subchapter, we will delve into the topic of addressing vitamin and mineral deficiencies in the context of GM food.

Vitamins and minerals play a crucial role in maintaining our overall health and well-being. However, deficiencies in these essential nutrients are prevalent, particularly in developing countries. This is where genetically modified food can potentially make a significant difference.

GM food has the potential to combat nutritional deficiencies by incorporating genes that enhance the nutrient content of crops. For instance, scientists have successfully developed GM crops that are fortified with essential vitamins such as vitamin A, also known as "golden rice." Vitamin A deficiency is a major public health issue, particularly in regions where rice is a staple food. By introducing genes responsible for vitamin A production into rice, scientists have created a biofortified crop that can address this deficiency.

Similarly, GM crops can be engineered to have increased levels of minerals such as iron and zinc. Iron deficiency, also known as anemia, affects millions of people worldwide, particularly women and children. By developing GM crops with higher iron content, we can potentially alleviate this health issue.

Addressing vitamin and mineral deficiencies through GM food is a promising solution; however, it is essential to consider the ethical, environmental, and economic implications. The public must be adequately informed about the benefits and potential risks associated with GM food. Regulation and labeling of GM products play a crucial role in ensuring transparency and consumer choice.

Moreover, it is important to assess the impact of GM crops on biodiversity. While GM crops may offer solutions to nutritional deficiencies, they must not harm the natural environment or contribute to the loss of biodiversity.

In conclusion, addressing vitamin and mineral deficiencies through genetically modified food has the potential to make a significant positive impact on global health. By creating biofortified crops, we can combat deficiencies in essential nutrients and improve the overall well-being of populations, particularly in developing countries. However, it is crucial to approach this technology with caution, considering the ethical, environmental, and economic implications associated with GM food. Regulation, labeling, and thorough scientific research are vital in ensuring the safe and responsible use of genetically modified crops to address nutritional deficiencies.

Implications for Malnutrition and Public Health

Malnutrition is a pressing global issue that affects millions of people, particularly those living in developing countries. In recent years, the role of genetically modified (GM) food in addressing malnutrition and improving public health has been a topic of debate. This subchapter aims to shed light on the implications of GM food on malnutrition and public health.

One potential benefit of GM food is its ability to combat nutritional deficiencies. Through genetic modification, crops can be enhanced to contain higher levels of essential nutrients such as vitamins and minerals. For example, scientists have developed GM rice known as "Golden Rice," which is fortified with beta-carotene, a precursor of vitamin A. This innovation holds significant promise in addressing vitamin A deficiency, a leading cause of blindness and immune system deficiencies in many developing countries.

Furthermore, the economic implications of GM food in developing countries cannot be ignored. By improving crop yields and reducing losses due to pests and diseases, GM crops can enhance food security and alleviate poverty. This, in turn, can contribute to better nutrition and overall public health outcomes. However, it is important to ensure that the benefits of GM food are accessible to smallholder farmers and do not exacerbate existing inequalities.

Nevertheless, concerns have been raised regarding the potential environmental impact of GM food production. Critics argue that genetically modified crops may harm biodiversity and disrupt ecosystems. Therefore, it is crucial to carefully evaluate the environmental consequences of GM food production and implement appropriate regulations to minimize any adverse effects.

Similarly, ethical considerations surrounding GM food cannot be overlooked. Some individuals and organizations raise concerns about the long-term health effects of consuming genetically modified organisms (GMOs). As such, it is essential to conduct rigorous scientific studies and ensure transparent labeling to allow consumers to make informed choices about the food they consume.

Another aspect to consider is the potential connection between GM food and food allergies. While there is currently no conclusive evidence linking GMO consumption to allergies, it is important to closely monitor and regulate the introduction of new GM crops to minimize any potential risks.

In summary, genetically modified food holds great potential in addressing malnutrition and improving public health. However, it is essential to consider the economic, environmental, ethical, and regulatory implications associated with GM food production. Through careful evaluation and monitoring, we can harness the benefits of GM food to combat malnutrition while safeguarding public health and the environment.

Chapter 8: The Impact of Genetically Modified Food on Biodiversity

Threats to Wild Plant and Animal Species

As the popularity and production of genetically modified (GM) food continues to increase, there are growing concerns about the potential threats it poses to wild plant and animal species. This subchapter aims to shed light on the various challenges that GM food production presents to biodiversity.

One of the primary concerns is the potential for gene flow between GM crops and their wild relatives. This occurs when GM crops cross-pollinate with wild plants, leading to the transfer of modified genes into the wild population. This can have significant consequences, such as the alteration of natural traits, reduced genetic diversity, and the potential for the creation of invasive species. These changes can disrupt ecosystems and pose a threat to the survival of native plant species.

Furthermore, the cultivation of GM crops often involves the use of pesticides and herbicides, which can have negative impacts on wildlife. These chemicals can contaminate water sources, disrupt food chains, and harm non-target organisms. For example, studies have shown that the widespread use of herbicides like glyphosate, commonly used in GM crop production, has led to a decline in milkweed plants, which are vital for the survival of monarch butterflies.

The expansion of GM crop cultivation also places increased pressure on natural habitats. As more land is converted for agricultural purposes, wildlife habitats are destroyed, leading to the loss of biodiversity. This habitat destruction affects not only wild plant species but also animals that rely on these habitats for food and shelter. The displacement of native species by GM crops can disrupt entire ecosystems and have far-reaching consequences.

It is crucial to address these threats and find ways to mitigate their impact. This requires robust regulation and monitoring of GM crop cultivation, as well as the development of sustainable farming practices. Additionally, more research is needed to understand the long-term effects of GM crops on biodiversity and ecosystem health.

In conclusion, the production of GM food presents significant threats to wild plant and animal species. Gene flow, pesticide use, habitat destruction, and the displacement of native species are just some of the challenges that need to be addressed. By understanding and addressing these threats, we can strive towards a more sustainable and balanced approach to food production that considers the well-being of both humans and the natural world.

Gene Flow and Hybridization

In the world of genetics, gene flow and hybridization play a significant role in the development and spread of genetically modified food. Gene flow refers to the movement of genes from one population to another, either through natural means like wind or insects, or through deliberate human intervention. Hybridization, on the other hand, occurs when two genetically distinct populations interbreed, resulting in offspring that possess a combination of traits from both parent populations.

In the context of genetically modified food, gene flow and hybridization can have both positive and negative implications. On one hand, gene flow can lead to the spread of desirable traits, such as increased yield or resistance to pests, from genetically modified crops to non-modified crops. This can potentially benefit farmers by improving the productivity and resilience of their crops. However, it can also raise concerns about the unintended consequences of genetic modification, as the spread of modified genes may have unforeseen effects on ecosystems and biodiversity.

Hybridization is another important aspect to consider. When genetically modified crops hybridize with non-modified crops, the resulting offspring may possess a mix of traits, some of which may be undesirable. For example, if a genetically modified crop is engineered to be herbicide-resistant and it hybridizes with a non-modified crop, the offspring may inherit the ability to survive herbicide exposure, potentially leading to the proliferation of herbicide-resistant weeds. This can have detrimental effects on agricultural practices and may require increased use of herbicides, leading to environmental concerns and potential health risks.

Understanding the potential impacts of gene flow and hybridization is crucial for assessing the safety and sustainability of genetically modified food. It raises important questions about the regulation and labeling of genetically modified crops, as well as the ethical considerations surrounding their production and consumption. Additionally, it highlights the need for rigorous testing and monitoring to ensure that genetically modified crops do not harm biodiversity or contribute to the development of new pests and diseases.

As public awareness and scrutiny of genetically modified food continue to grow, it is essential for individuals to understand the complexities of gene flow and hybridization. By gaining insight into these concepts, the public can make informed decisions about the health effects, environmental impact, ethical considerations, and economic implications of genetically modified food. Furthermore, further research and development in this area can help harness the potential of genetically modified food to combat nutritional deficiencies and improve global food security while minimizing the risks to biodiversity and human health.

Conservation and Preservation Efforts

In the pursuit of addressing the various concerns surrounding genetically modified (GM) food, it is crucial to consider the conservation and preservation efforts necessary to mitigate potential risks. These efforts encompass a broad range of areas including environmental sustainability, biodiversity conservation, ethical considerations, and long-term economic implications.

One of the key concerns regarding GM food is its potential negative impact on the environment. The production of GM crops often involves the use of intensive agricultural practices that can lead to soil degradation, water contamination, and loss of biodiversity. To address these concerns, conservationists and environmentalists advocate for the adoption of sustainable farming practices, such as organic farming and agroecology, which promote soil health, reduce chemical inputs, and preserve ecological balance. By embracing these practices, we can minimize the environmental footprint of GM food production and ensure the long-term health of our ecosystems.

Another important aspect of conservation and preservation efforts is the need to protect biodiversity. GM crops have the potential to crossbreed with wild relatives, leading to the spread of modified genes into natural habitats. This poses a risk to indigenous plant species and can disrupt delicate ecosystems. To mitigate this risk, strict regulations and monitoring systems should be in place to prevent the unintended spread of GM traits. Additionally, efforts should be made to conserve and protect natural habitats that serve as repositories of genetic diversity, providing a vital resource for future crop improvement and adaptation to changing environmental conditions.

Ethical considerations also come into play when discussing GM food. The potential for corporate control over the seed market and intellectual property rights raises concerns about equitable access to food resources, particularly in developing countries. Conservation and preservation efforts should therefore focus on promoting fair and sustainable agricultural systems that prioritize the needs of small-scale farmers and local communities, ensuring food security and economic empowerment.

Furthermore, conservation and preservation efforts should not only consider the short-term economic implications but also the long-term sustainability of GM food production. It is essential to evaluate the economic viability of GM crops in developing countries, considering factors such as market demand, access to technology, and potential impacts on local agriculture. By conducting thorough economic assessments, policymakers can make informed decisions on the adoption and regulation of GM food, ensuring both economic growth and environmental sustainability.

In conclusion, conservation and preservation efforts are crucial when discussing the truth about genetically modified food. By focusing on sustainable agricultural practices, biodiversity conservation, ethical considerations, and long-term economic implications, we can better address the concerns surrounding GM food and ensure a more sustainable future for our food systems.

Chapter 9: The Regulation and Labeling of Genetically Modified Food

International Regulatory Frameworks

In today's interconnected world, where goods and information flow across borders effortlessly, it is crucial to have international regulatory frameworks in place to govern various aspects of global trade. This is particularly true when it comes to controversial topics such as genetically modified (GM) food. The regulation and labeling of genetically modified food have become significant concerns for numerous stakeholders, including the public, health experts, environmentalists, ethicists, economists, and developing countries.

The truth about genetically modified food is a highly debated topic. Many people have concerns about the health effects of consuming GMOs, especially when it comes to children. They worry about potential allergenic reactions, long-term health consequences, and the unknown risks associated with altering the genetic makeup of our food. To address these concerns, international regulatory frameworks play a crucial role in ensuring that GM products undergo rigorous testing and evaluation before they are approved for consumption.

Moreover, the environmental impact of genetically modified food production cannot be ignored. The use of GM crops often involves the application of pesticides and herbicides, which can have adverse effects on ecosystems and biodiversity. Regulatory frameworks must address these concerns by setting standards for sustainable farming practices, promoting crop rotation, and encouraging the use of alternative pest management techniques.

Ethical considerations also come into play when discussing genetically modified food. Some argue that tampering with the genetic makeup of organisms goes against the natural order of things and interferes with the concept of food as a basic human right. International regulatory frameworks must weigh these ethical concerns against the potential benefits of GM food, such as increased crop yields and enhanced nutritional value.

The economic implications of genetically modified food in developing countries are also significant. While GM crops may offer increased productivity and improved resistance to pests and diseases, they can also create dependencies on multinational corporations and restrict farmers' access to traditional seeds. Regulatory frameworks must address these concerns by promoting fair trade practices and protecting the rights of small-scale farmers.

Additionally, international regulatory frameworks play a crucial role in ensuring global food security. Genetically modified food has the potential to combat nutritional deficiencies by enhancing the nutritional content of staple crops. However, the distribution and affordability of these genetically modified crops must be carefully regulated to ensure equitable access for all populations, especially in developing countries.

In conclusion, international regulatory frameworks are essential for addressing the various concerns surrounding genetically modified food. By regulating and labeling GM products, addressing health effects, environmental impact, ethical considerations, economic implications, and promoting global food security, these frameworks provide a foundation for informed decision-making and responsible governance in the era of genetically modified food.

Labeling Laws and Consumer Rights

The regulation and labeling of genetically modified food is a topic of utmost importance when it comes to consumer rights. Consumers have the right to know what they are purchasing and consuming, and labeling laws play a crucial role in ensuring this transparency. In this subchapter, we will delve into the various aspects related to labeling laws and their significance in protecting consumer rights.

Labeling laws vary across different countries and jurisdictions, but their main objective is to provide consumers with accurate and comprehensive information about the presence of genetically modified organisms (GMOs) in food products. These laws often require that GMO-containing products be clearly labeled, allowing consumers to make informed choices based on their personal preferences, health concerns, or ethical considerations.

For those concerned about the health effects of genetically modified food on children, labeling laws can provide a sense of control and peace of mind. Parents can easily identify and avoid GMO-containing products if they wish to do so, safeguarding their children's health and well-being.

Furthermore, labeling laws also address the environmental impact of genetically modified food production. By including information about GMOs on food labels, consumers can support sustainable agriculture and make choices that align with their values. This empowers individuals to take an active role in preserving biodiversity and promoting environmentally friendly practices.

Ethical considerations surrounding genetically modified food are also taken into account through labeling laws. Consumers who have moral concerns about the use of GMOs can opt for non-GMO products, supporting businesses that align with their ethical beliefs.

Additionally, labeling laws have economic implications, particularly in developing countries. By requiring proper labeling, these laws promote transparency in the global food trade and help protect local farmers from unfair competition. Developing nations can also use labeling as a tool to differentiate their products in the international market, potentially improving their economic standing.

Ultimately, labeling laws are crucial for protecting consumer rights and ensuring transparency in the food industry. They empower individuals to make informed choices based on their own values, health concerns, and environmental considerations. By understanding and supporting labeling laws, we can collectively advocate for consumer rights and contribute to a more sustainable and equitable food system.

Transparency and Disclosure

In the world of genetically modified (GM) food, transparency and disclosure play a crucial role in ensuring public trust and understanding. The public has the right to know what they are consuming and the potential impacts it may have on their health, the environment, and society as a whole. This subchapter will delve into the importance of transparency and disclosure in the context of GM food, addressing the concerns and interests of various niches within the public.

When it comes to the truth about genetically modified food, transparency is key. Clear and accessible information should be provided to the public regarding the development, testing, and regulation of GM crops. This transparency allows individuals to make informed decisions about what they choose to eat, empowering them to protect their health and well-being.

Speaking of health effects, children are a particularly vulnerable group. The subchapter will shed light on the potential health impacts of genetically modified food on children. It will explore scientific studies and evidence, discussing the safety of GM foods in relation to children's health and development.

Furthermore, the environmental impact of GM food production cannot be overlooked. The subchapter will examine the effects of GM crops on ecosystems, biodiversity, and the use of natural resources. It will also address concerns about the long-term consequences of genetic modification on the environment and sustainable agricultural practices.

Ethical considerations surrounding GM food will also be explored. This includes the moral implications of genetic modification, such as the ownership of seeds and the potential exploitation of farmers in developing countries. The subchapter will analyze the ethical dilemmas posed by the commercialization of GM crops and the potential implications for food sovereignty.

Moreover, the economic implications of GM food in developing countries will be discussed. The subchapter will examine how the adoption of GM crops affects small-scale farmers, local economies, and food security. It will evaluate the impact of intellectual property rights and the concentration of power in the hands of a few multinational corporations.

In terms of global food security, genetically modified food has the potential to address nutritional deficiencies. The subchapter will explore the role of GM crops in providing enhanced nutrition, such as biofortification, and the potential benefits for developing countries in combating malnutrition.

The impact of GM food on biodiversity is another crucial aspect to consider. The subchapter will examine the potential risks and benefits of genetic modification on biodiversity, including the potential loss of traditional crop varieties and the unintended consequences on ecosystems.

Regulation and labeling of GM food will also be addressed. The subchapter will explore the current regulatory frameworks and labeling practices in different countries, discussing the need for standardized regulations and clear labeling to allow consumers to make informed choices.

Furthermore, the connection between genetically modified food and food allergies will be explored. The subchapter will discuss scientific evidence and studies on the potential allergenicity of GM crops and the importance of labeling for individuals with food allergies.

Lastly, the subchapter will delve into the development of genetically modified crops resistant to pests and diseases. It will examine the benefits and risks associated with this approach, including the potential reduction in pesticide use and the emergence of resistant pests.

In conclusion, transparency and disclosure are vital components in the discussions surrounding genetically modified food. This subchapter aims to provide the public with a comprehensive understanding of various aspects related to GM food, addressing the concerns and interests of different niches within the public. By promoting transparency and disclosure, individuals can make informed decisions about GM food and its potential impacts on health, the environment, ethics, economics, global food security, biodiversity, regulation, labeling, food allergies, and pest resistance.

Chapter 10: Genetically Modified Food and Its Connection to Food Allergies

Allergenic Potential of Genetically Modified Ingredients

Genetically modified (GM) ingredients have become increasingly prevalent in our food supply, raising concerns about their potential allergenicity. While proponents argue that GM foods are safe for consumption, critics question the long-term effects of consuming these altered ingredients. This subchapter will explore the allergenic potential of genetically modified ingredients, shedding light on the potential risks they may pose to human health.

Allergies are a growing concern worldwide, affecting millions of people. Some individuals are allergic to specific foods, such as peanuts or shellfish, while others may have sensitivities to certain ingredients. The introduction of genetically modified ingredients into the food chain has raised questions about whether these alterations could trigger allergies or exacerbate existing sensitivities.

One of the primary concerns is the transfer of allergenic proteins from genetically modified crops to the final food products. Genetic modification involves the insertion of foreign genes into plants, often from unrelated species. While the intention is to enhance desired traits, there is a possibility that these inserted genes could encode proteins that are allergenic to some individuals.

To address this concern, regulatory bodies in many countries have established guidelines for assessing the allergenic potential of genetically modified ingredients. These guidelines typically involve evaluating the source of the inserted gene, its similarity to known allergens, and the stability of the protein produced. Additionally, animal testing is often conducted to assess the potential allergenicity of genetically modified ingredients.

Despite these precautions, there have been instances where genetically modified ingredients have been found to be allergenic. For example, in the late 1990s, a genetically modified soybean variety was withdrawn from the market after it was discovered to cause allergic reactions in individuals sensitive to Brazil nuts, to which the inserted gene was related.

To further complicate matters, the potential allergenicity of genetically modified ingredients may also depend on the specific genetic modification technique used. For instance, some studies suggest that genetic modifications that involve the use of RNA interference could result in the production of small RNA molecules that could be allergenic to certain individuals.

In conclusion, the allergenic potential of genetically modified ingredients is a significant concern. While regulatory measures are in place to assess and mitigate this risk, there have been instances where genetically modified ingredients have been found to be allergenic. As the consumption of genetically modified foods continues to rise, further research and vigilance are needed to ensure the safety of these ingredients for individuals with food allergies or sensitivities.

Testing and Detection Methods

In order to understand the truth about genetically modified food, it is crucial to be aware of the various testing and detection methods used to identify genetically modified organisms (GMOs) in our food supply. These methods play a vital role in ensuring transparency and providing consumers with accurate information about the presence of GMOs in the products they consume.

One commonly used testing method is polymerase chain reaction (PCR). PCR allows scientists to amplify and detect specific DNA sequences, enabling them to identify the presence of genetically modified ingredients. This method has proven to be highly accurate and reliable in detecting GMOs, and it is widely used by regulatory authorities and independent laboratories.

Another testing method is enzyme-linked immunosorbent assay (ELISA). ELISA uses antibodies to detect specific proteins present in GMOs. This method is particularly useful for detecting genetically modified proteins, such as the Bt toxin found in some GM crops. ELISA is a cost-effective and relatively quick method, making it widely used in the food industry.

In addition to these laboratory-based methods, there are also field-based testing methods available. One such method is lateral flow strips or dipsticks, which are similar to pregnancy test kits. These strips can detect the presence of GMOs through a color change reaction, providing a rapid and on-site detection method that does not require specialized equipment.

It is important to note that testing methods alone are not enough to ensure the safety and transparency of genetically modified food. The development and implementation of robust regulations and labeling requirements are equally crucial. Governments and regulatory bodies play a significant role in monitoring and enforcing these regulations to protect consumer rights.

By understanding the testing and detection methods used to identify GMOs, consumers can make informed choices about the food they consume. This knowledge empowers individuals to exercise their right to know and make decisions that align with their values and concerns.

In conclusion, testing and detection methods are essential tools in uncovering the truth about genetically modified food. They provide accurate information about the presence of GMOs, enabling consumers to make informed choices. Robust regulations and labeling requirements, along with these testing methods, ensure transparency and protect consumer rights. By staying informed, individuals can actively contribute to the ongoing dialogue surrounding genetically modified food and its implications for health, the environment, ethics, and global food security.

Prevalence and Management of Allergic Reactions

Allergic reactions to food are a growing concern among the public, especially in relation to genetically modified (GM) food. In this subchapter, we will explore the prevalence of allergic reactions to GM food and discuss strategies for managing these reactions.

The Truth About Genetically Modified Food has sparked intense debate and raised questions about the safety of consuming GM food. One major concern is the potential for these foods to trigger allergies. It is important to note that allergic reactions can occur with any type of food, GM or non-GM, and they vary in severity from mild to life-threatening.

Research suggests that the prevalence of food allergies, including those related to GM food, is increasing globally. Children are particularly vulnerable to allergic reactions, and there is a growing body of evidence linking the consumption of GM food to adverse health effects in children. It is crucial for parents and caregivers to be aware of these risks and take necessary precautions.

Managing allergic reactions to GM food involves a multi-faceted approach. Firstly, individuals with known allergies should carefully read food labels to identify any potential GM ingredients. However, it is worth mentioning that regulations regarding the labeling of GM food vary across countries, making it challenging for consumers to make informed choices.

Furthermore, individuals who suspect they have developed allergies to GM food should consult healthcare professionals for proper diagnosis and guidance. Allergy testing can help identify specific allergens, enabling individuals to avoid them and minimize the risk of reactions.

In addition to personal management, it is essential for regulatory bodies to establish and enforce strict guidelines for the production and labeling of GM food. This would enable consumers to make informed decisions about the food they purchase and consume, especially for those with existing allergies or a family history of allergic reactions.

Furthermore, ongoing research is necessary to better understand the connection between GM food and allergies. This will help in developing improved diagnostic tools, treatment options, and preventive measures.

In conclusion, allergic reactions to GM food are a significant concern that needs to be addressed. The public, particularly those interested in the truth about genetically modified food, the health effects on children, the environmental impact, and ethical considerations, should be aware of the prevalence and management of these reactions. By promoting transparency, investing in research, and implementing effective regulations, we can strive towards a safer and healthier future for all consumers.

Chapter 11: Genetically Modified Crops and Their Resistance to Pests and Diseases

Integrated Pest Management Strategies

Integrated Pest Management (IPM) is a holistic approach to pest control that focuses on minimizing the use of synthetic pesticides and promoting environmentally friendly alternatives. In the context of genetically modified (GM) food production, IPM strategies play a crucial role in reducing the reliance on chemical pesticides and ensuring sustainable agricultural practices. This subchapter explores the various IPM strategies employed in GM food production, their benefits, and their implications for the public, children's health, the environment, ethics, economies, global food security, nutrition, biodiversity, regulation, labeling, and pest and disease resistance.

IPM strategies involve a combination of techniques to manage pests effectively while minimizing their impact on the environment and human health. These techniques include biological control, cultural practices, crop rotation, habitat manipulation, and the use of resistant crop varieties. By incorporating these strategies, GM food production can reduce the need for chemical pesticides, thereby decreasing the potential risks associated with their use.

From a public health perspective, IPM strategies in GM food production can minimize exposure to harmful pesticides, reducing the risk of pesticide residues in food. This is particularly important for children, as they are more vulnerable to the potential health effects of pesticide exposure. By implementing IPM practices, GM foods can offer a safer and healthier option for children's diets.

Moreover, IPM strategies contribute to the preservation of the environment by promoting biodiversity and reducing the negative impacts of pesticide use on ecosystems. By encouraging the use of natural predators and beneficial insects, GM crops can help maintain a balanced ecosystem and reduce the need for chemical interventions.

Ethically, IPM practices align with the principles of sustainability and responsible stewardship of the environment. By adopting these strategies, GM food production demonstrates a commitment to minimizing harm and maximizing the long-term benefits for both current and future generations.

Economically, IPM strategies can benefit developing countries by reducing the reliance on costly chemical pesticides and promoting sustainable agricultural practices. This can lead to increased productivity, improved livelihoods, and economic stability in these regions.

In terms of global food security, IPM strategies can contribute by improving crop yields, reducing post-harvest losses, and promoting sustainable farming practices. By minimizing the impact of pests and diseases, GM crops can help ensure a more reliable and abundant food supply for a growing global population.

Furthermore, IPM strategies can be utilized to address nutritional deficiencies through the development of GM crops with enhanced nutritional profiles. By incorporating essential vitamins or minerals into crops, GM food production can play a crucial role in combating malnutrition and improving public health.

In conclusion, integrated pest management strategies in GM food production offer a comprehensive and sustainable approach to pest control. These strategies have far-reaching implications for the public, children's health, the environment, ethics, economies, global food security, nutrition, biodiversity, regulation, labeling, and pest and disease resistance. By adopting IPM practices, the production of genetically modified food can contribute to a safer, healthier, and more sustainable food system for all.

Reduction of Chemical Pesticide Use

One of the key concerns surrounding genetically modified food (GM food) is the heavy reliance on chemical pesticides in its production. However, there is a growing movement towards reducing the use of these harmful chemicals and finding alternative methods to protect crops from pests and diseases.

The use of chemical pesticides in conventional agriculture has been linked to a range of health issues, including respiratory problems, neurodevelopmental disorders, and even certain types of cancer. This is particularly worrying when it comes to children, as they are more vulnerable to the harmful effects of these chemicals. Therefore, it is essential to explore ways to reduce chemical pesticide use in GM food production to protect the health of our children.

One of the most promising solutions is the development of genetically modified crops that are resistant to pests and diseases. By introducing specific genes into these crops, scientists have been able to enhance their natural defense mechanisms, reducing the need for chemical pesticides. For example, Bt crops, which produce their own insecticides, have shown great success in reducing the use of chemical sprays.

Additionally, integrated pest management (IPM) practices can also play a significant role in reducing chemical pesticide use. IPM involves a combination of techniques such as crop rotation, biological control, and the use of pheromones to disrupt pest mating patterns. This holistic approach not only reduces the reliance on chemical pesticides but also promotes biodiversity and overall ecosystem health.

Furthermore, promoting organic farming methods can also contribute to the reduction of chemical pesticide use. Organic farming relies on natural pest control methods, such as beneficial insects and crop diversification, rather than synthetic chemicals. By supporting organic agriculture and encouraging farmers to adopt organic practices, we can create a healthier and more sustainable food system.

In conclusion, reducing the use of chemical pesticides in GM food production is crucial for the health of our children, the environment, and our overall food system. Through the development of genetically modified crops, the implementation of integrated pest management practices, and the promotion of organic farming methods, we can minimize the reliance on harmful chemicals and pave the way for a safer and more sustainable future.

Implications for Sustainable Agriculture

Sustainable agriculture is a holistic approach to farming that aims to meet the growing needs of the present generation without compromising the ability of future generations to meet their own needs. It emphasizes the use of environmentally friendly farming practices, protection of natural resources, and the promotion of food security. When it comes to genetically modified food, there are several implications for sustainable agriculture that need to be considered.

One of the key benefits of genetically modified (GM) crops in sustainable agriculture is their potential to reduce the use of chemical pesticides and herbicides. GM crops can be engineered to resist pests and diseases, allowing farmers to minimize their reliance on harmful chemicals. This not only reduces the environmental impact of agriculture but also improves the safety of food for consumers.

Additionally, GM crops have the potential to increase crop yields, which can help meet the growing demand for food in a sustainable manner. By incorporating traits like drought resistance and increased nutritional content, GM crops can thrive in challenging environmental conditions and provide more nutritious food for communities. This is especially important in developing countries, where food security is a pressing issue.

However, it is essential to carefully consider the potential risks and ethical implications associated with genetically modified food. The environmental impact of GM crop production needs to be assessed to ensure that it does not harm biodiversity or ecosystems. The regulation and labeling of GM food are also crucial to provide consumers with the information they need to make informed choices about the food they consume.

Furthermore, the potential health effects of genetically modified food, particularly on children, need to be thoroughly studied and understood. Research should focus on the long-term impacts and potential allergenicity of GM crops to protect the health of vulnerable populations.

In conclusion, genetically modified food has both opportunities and challenges for sustainable agriculture. While it has the potential to increase crop yields, reduce chemical pesticide use, and improve food security, it also raises concerns about environmental impact, health effects, and ethical considerations. It is essential that policymakers, scientists, and the public work together to ensure that genetically modified food is carefully regulated, labeled, and researched to maximize its benefits and minimize its potential drawbacks in sustainable agriculture.

Conclusion: Navigating the Complexities of Genetically Modified Food

In this book, "Unveiling the Secrets: The Truth About Genetically Modified Food," we have delved into the multifaceted world of genetically modified food (GMF) and explored its various dimensions. Throughout our journey, we have addressed the concerns and queries of the public, as well as the niches of The Truth About Genetically Modified Food, the health effects of genetically modified food on children, the environmental impact of genetically modified food production, the ethical considerations of genetically modified food, the economic implications of genetically modified food in developing countries, the role of genetically modified food in global food security, genetically modified food and its potential to combat nutritional deficiencies, the impact of genetically modified food on biodiversity, the regulation and labeling of genetically modified food, genetically modified food and its connection to food allergies, and genetically modified crops and their resistance to pests and diseases.

Through analyzing various scientific studies and expert opinions, we have gained a comprehensive understanding of the benefits and risks associated with GMF. It is crucial to acknowledge that GMF has the potential to address some of the pressing challenges faced by our planet, such as food scarcity, malnutrition, and environmental degradation. However, we must also recognize the need for responsible and transparent practices to ensure the safety and sustainability of genetically modified crops.

One of the key takeaways from our exploration is the importance of informed decision-making. It is essential for the public to have access to accurate and unbiased information about GMF, enabling them to make educated choices about the food they consume. Furthermore, policymakers and regulatory bodies must establish robust frameworks for the assessment, labeling, and monitoring of GMF to safeguard public health and environmental integrity.

While the debate surrounding GMF will continue, it is crucial to foster dialogue and collaboration among stakeholders from diverse backgrounds. By considering the perspectives of scientists, farmers, consumers, and environmentalists, we can navigate the complexities of GMF and work towards sustainable solutions.

In conclusion, the realm of genetically modified food is complex and encompasses a wide range of issues. As we move forward, it is vital to approach this topic with an open mind, acknowledging both the potential benefits and risks of GMF. By fostering transparency, responsible practices, and informed decision-making, we can ensure that genetically modified food plays a constructive role in addressing global challenges while protecting human health, environmental integrity, and biodiversity.